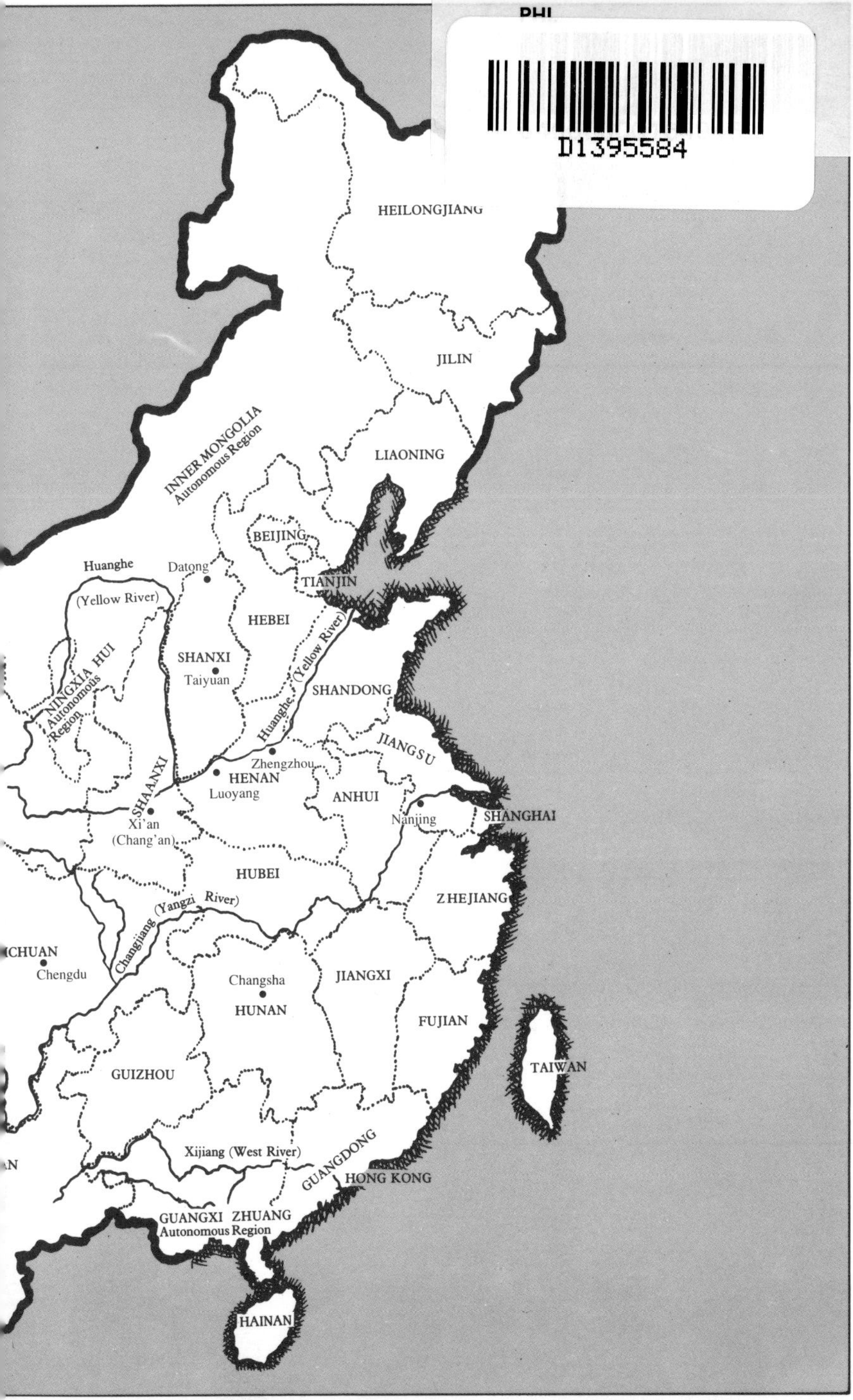
HEILONGJIANG
JILIN
LIAONING
INNER MONGOLIA
Autonomous Region
BEIJING
TIANJIN
Huanghe
(Yellow River)
Datong
HEBEI
SHANXI
Taiyuan
NINGXIA HUI
Autonomous
Region
Huanghe (Yellow River)
SHANDONG
JIANGSU
Zhengzhou
HENAN
Luoyang
SHAANXI
Xi'an
(Chang'an)
ANHUI
Nanjing
SHANGHAI
HUBEI
Changjiang (Yangzi River)
ZHEJIANG
Chengdu
Changsha
HUNAN
JIANGXI
FUJIAN
TAIWAN
GUIZHOU
Xijiang (West River)
GUANGDONG
HONG KONG
GUANGXI ZHUANG
Autonomous Region
HAINAN

IMAGES OF ASIA

Chinese Tomb Figurines

Series Editors, China Titles:
NIGEL CAMERON, SYLVIA FRASER-LU

Chinese Tomb Figurines

ANN PALUDAN

HONG KONG
OXFORD UNIVERSITY PRESS
OXFORD NEW YORK
1994

Oxford University Press

Oxford   New York   Toronto
Kuala Lumpur   Singapore   Hong Kong   Tokyo
Delhi   Bombay   Calcutta   Madras   Karachi
Nairobi   Dar es Salaam   Cape Town
Melbourne   Auckland   Madrid

and associated companies in
Berlin   Ibadan

Oxford is a trade mark of Oxford University Press

First published 1994

Published in the United States
by Oxford University Press, New York

© Oxford University Press 1994

British Library Cataloguing in Publication Data available

Library of Congress Cataloging-in-Publication Data
Paludan, Ann, 1928–
Chinese tomb figurines / Ann Paludan.
p.   cm. — (Images of Asia)
Includes bibliographical references and index.
ISBN 0–19–585817–4
1. Ming ch'i.   2. Funeral rites and ceremonies—China.   I. Title.
II. Series.
NK4165.P28   1994
732′.71—dc20   93–44794
                                                    CIP

Printed in Hong Kong
Published by Oxford University Press (Hong Kong) Ltd
18/F Warwick House, Taikoo Place, 979 King's Road, Quarry Bay,
Hong Kong

Contents

Acknowledgements

I WISH to thank the Royal Ontario Museum, Toronto, especially Jeannie Parker and Patty Proctor for their kind welcome and their practical and professional assistance.

I also thank the following for their generous help and advice: Christina Hsu, The Art Museum, Princeton; Rose Kerr, Victoria and Albert Museum, London; Ladislav Kesner, The National Gallery, Prague; Candace Lewis, Vassar College, Poughkeepsie, N.Y.; Margaret Medley; Nicholas Pearce, The Burrell Collection, Glasgow; Jessica Rawson, Jessica Harrison-Hall, British Museum, London; Sheilagh Vainker, Ashmolean Museum, Oxford; Dr. Frances Wood, British Library, London. Above all, I wish to thank my husband, Janus, for his help and support. It was he who convinced me many years ago that tomb figurines were an important element in the study of Chinese sculpture.

Introduction

THE wish to comfort the dead lies deep in mankind. Small pots or household utensils, decorative objects like beads, or useful tools such as knives, arrowheads, or fish hooks have been placed in primitive tombs since earliest times. In more advanced societies, the dead were provided with replicas of worldly goods or talismans to help them in their new life. In the fourteenth century BC, the young Egyptian king, Tutankhamen, was buried with an exquisite model boat and numerous statuettes of himself enabling him to continue to perform his different functions after death, whilst the Greeks in the sixth and fifth centuries BC were provided with model figures on horseback and miniature jars to keep their precious oil. None of these civilizations, however, approaches the scope and continuity of the Chinese tomb figurine tradition. In China, such burial objects became an integral part of the all-important tomb system and were used to recreate the entire world of the living within the tomb.

Wandering into the world of Chinese figurines is like stepping into a time warp. Amongst figurines of the great Han Dynasty (206 BC–AD 220), you find yourself in the countryside. A farmer rests on his spade while a sow in the pigsty suckles her young. In a large house the dog keeps watch in the porter's lodge, hens roost on the granary roof, small bronze pots and pans stand ready in the kitchen. Elsewhere a cook vigorously scrapes the scales from a fish while nearby in a tea house, guests sit chatting, entertained by a dwarf story-teller. Then, as if turning the handle of an old movie projector, time moves on. Northern China has been overrun by nomadic horsemen from the steppes, and suddenly you are whisked from domestic agricultural life to face bearded warriors with non-Chinese features, booted horsemen with hooked noses sheltered from the bitter wind of the steppes by long coats hung nonchalantly over their shoulders and warm balaclava helmets. Hunters on high-spirited horses pass heavily laden camels; musicians including lady drummers escort dignitaries carried in ox-drawn carts.

The handle turns and once again the scene changes. This is seventh and eighth century China — time of the great Tang Empire, often known as China's Golden Age. Here is a world of elegant luxury, of aristocratic pursuits such as polo and falconry, of fashion with highly made-up beauties in tight-fitting dresses with daring necklines or loose diaphanous robes, stoles or sashes blowing in the wind as they dance. Here are horses of every ilk and hue — reflecting a world in which, as we know from the historical records, the emperor taught his favourite steeds to dance and drink wine from a golden cup. A last glimpse of this moving panorama shows the exquisite furniture of the sixteenth century — folding chairs, low tables, bookshelves, and cupboards, courtyard houses and funeral processions with musicians accompanying the deceased on his last earthly journey.

Tens of thousands of these figures have been unearthed in China in the last few decades, and each newly excavated tomb adds to the score. The vast quantity of surviving figurines is partly a reflection of the sheer number produced in earlier periods and partly because these small clay figures were ignored by grave robbers in search of bronzes, jades, and other intrinsically valuable objects. Baked clay is a durable material, impervious to damp and rot, and unless broken, the figurines have survived in their original condition with only the colouring impaired.

Recognition of the value of these little models did not come until the twentieth century. One of the earliest to recognize their worth was a Beijing collector and art connoisseur, Luo Zhenyu, to whom some grave robbers showed a couple of ceramic figures which they had picked up on leaving a tomb. Recognizing their historical interest, Luo Zhenyu asked them to find more and soon amassed a considerable collection, publishing a book about them in 1916. Early Western works, such as Berthold Laufer's 'Chinese Clay Figures' (1914), Carl Hentze's *Chinese Tomb Figures* (1928), and Jane Mahler's *Westerners among the Figurines of the T'ang Dynasty of China* (1959), concentrated on the value of these artefacts as a reflection of daily life in the past. Only recently have figurines attracted the interest of the art market. Today there are probably more figurines in private and museum collections in the West than any other form of Chinese artefact except porcelain, and individual examples have changed hands for over a million pounds.

Small figures of human beings and animals were made in many materials and for different purposes. The figurines with which this book is concerned are ceramic and occasionally wooden figures, made specifically for the tomb and designed primarily as substitutes for real objects. Their role is inextricably bound up with the tomb and the all-important Chinese practice of ancestor worship based on the belief that the spirits of the deceased exerted an active influence on the lives of their descendants. This book traces their history, explaining the ideas behind their use and the way in which they reflected contemporary conditions and beliefs. It examines the technical processes involved and describes the development whereby the use of a few clay imitations of valuable objects was transformed into a vast industry with official workshops mass-producing tens of thousands of models. Finally, it considers the role played by this living form of plastic art in the classical tradition of Chinese sculpture, a tradition which has its roots in prehistory but is still alive in the countryside today.

1

Origins

THE use of Chinese clay tomb figurines developed from the need to produce cheap substitutes for articles of value. In early Chinese history, there was little scope for substitution. Neolithic tombs were rarely large enough to allow for more than a few articles to accompany the dead. Some small clay models of human heads or complete figures including that of a pregnant woman as well as miniature clay animals and birds have been found in Neolithic tombs, but the nature of these subjects suggests that they were created primarily for religious or magical purposes. During the Bronze Age which followed, the association of the tomb with ancestor worship increased the importance of burial goods, and excavations of the large royal graves of the Shang (c. 1550–1027 BC) and Western Zhou (1027–771 BC) have revealed a wealth of tomb offerings, the most important of which were ritual bronze vessels used in the rites and sacrifices to contact the spirit world. Most of the earliest known clay substitutes are imitations of these bronzes found in lesser tombs whose occupants were presumably either unable to afford or not entitled to use the genuine article. Made of rough low-fired clay, they were decorated to simulate the original bronzes. Patterns were incised or impressed and painted with pigments or lacquer; more elaborate examples were inlaid with glass paste or tin or copper foil. The remaining figurines from this period — small clay and stone animals and birds such as tigers, turtles, fish, owls, and goat and ox heads — were apparently made as substitutes for the numerous and extremely valuable jade talismans placed in wealthy tombs.

For most of the Bronze Age, the wish to treat the dead

as if they were alive entailed large scale human sacrifice in order to provide the deceased with attendants. Around the fifth century BC, however, reaction against the cruelty and wastefulness of such a system led to the gradual introduction of substitute human figurines. By the Warring States period (475–221 BC) substitutes were in general use. In central China these were of clay, but in the southern Kingdom of Chu in modern Henan, wood and straw figurines were used (Fig. 1.1 and Plate 1). No straw models have survived, and some of the earliest wooden examples with long tongues and antlers from Chu tombs appear to have served magical rather than substitute purposes. Other wooden models, however, were made like puppets with moveable limbs. An example from a tomb in Laixi, Shandong, is 1.93 metres tall, with openings on the head for real hair; it has thirteen moveable parts which were originally assembled with tenons so that it could stand or kneel. When Confucius praised straw figurines but criticized wooden ones on the grounds that they might lead to a recurrence of human sacrifice, it seems likely that it was these puppet creatures he had in mind.

As long as tomb furnishings were primarily concerned with status and the correct performance of the rites, the scope for figurines remained limited. In the fourth century BC, however, a new theory about the nature of the soul and the way to communicate with the other world led to fundamental changes. When the body died, the soul was now thought to split into two: a 'spirit-soul', responsible for thought, creativity, and morality, which travelled to the land of the immortals, and an 'earth-soul', responsible for life and movement, which remained with the body in the tomb. Contact with the other world was no longer direct, through sacrifices, but indirect, through the family ancestors. If the spirit of the deceased were happy, it would

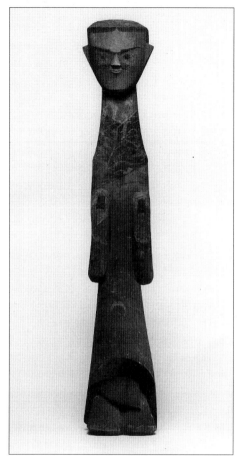

1.1 Painted wooden figure, Eastern Zhou period (771–256 BC). The arms, possibly made of a different material, were fastened with wooden pegs in holes in the sleeves. Height 56.6 cm. The Art Museum, Princeton University. Museum purchase, Carl Otto von Kienbusch, Jr. Memorial Collection.

intercede in the spirit world on behalf of its descendants and they would prosper; if it were neglected, it would take revenge and the whole family would suffer. Tomb furnishings had therefore to satisfy both parts of the soul. The spirit-soul was provided with talismans and nourishment from sacrifices for its journey into the unknown. Life in the underground world of the earth-soul was conceived as a continuation of life in this world, and the earth-soul had therefore to be provided with all the necessities and lux-uries to which it had been accustomed during life. From being a treasure chest of priceless ceremonial objects, the tomb was transformed into a dwelling.

The single pit tombs of the Bronze Age were now enlarged to include several chambers, sometimes decorated like a house. The desire to satisfy the earth-soul with familiar objects led to an increasing use of clay models. Whereas it had been possible to bury horses, chariots, bronze vessels, and even human beings, it was hardly practical to bury all the paraphernalia of daily life, and pottery models of ducks, geese, and other animals, carriages, kettles, and even gran-aries have been excavated from tombs of this period. One bird is made of six interlocking parts, but most are simple figures like a set of nine male and female dancers and musi-cians found in Shanxi. Only 5 cm high, their sketchily modelled bodies are little more than rough cylinders, but they show the characteristic Chinese balance between line and form, and the artist has caught the movement and vitality of the entertainers.

It was against this background that the First Emperor of China, Qin Shihuangdi (r. 221–210 BC), planned his tomb. His army of 7,000 life-size clay warriors was part of a grandiose attempt to recreate the known world below ground. According to Chinese histories, the tomb contained palaces and towers. The rivers of China were recreated in mercury, whilst the planets and stars were set in pearls in the cop-per domed roof. Outside the tomb, the picture of the real world was continued with figures buried in pits. Nearly one hundred kneeling clay figures, each beside a live horse burial, represented the imperial stables (Plate 2); similar figures beside urns with animal skeletons represented the imperial menagerie. In large pits to the south-east, com-pleting the picture of an empire based on military force, was the army. Even today, the creation of over 7,000 realistically portrayed life-size figures would be a major undertaking. With the transport means and technology then available it was a feat of monumental proportions, and the production methods developed provided a precedent for later figurine manufacture. The choice of clay as a medium appears to have been made for practical reasons. Clay was cheap, abundant, and easy to work, and there already existed a workforce with a tradition of over a thousand years of experience in using clay cast-moulds for bronzes. The organization of workshops for piece-work specialization and all the techniques for mass production which had been developed in the Bronze Age were now adapted to produce clay figures. Like the bronzes, these were hollow, made from separately moulded parts assembled and decorated before firing; like the early clay imitations of bronzes they were then painted after firing.

The First Emperor's army was a unique experiment. Its effects, however, were long lasting. The literal approach to the problem of satisfying the soul by recreating the conditions of this world for its existence in the next lay behind the great tomb extravagance of the Han Dynasty which followed. At the same time, the use of clay for an imperial project of this order transformed attitudes towards clay as a medium, extending its use beyond that of creating utilitarian or ceremonial vessels. Its advantages had been clearly demonstrated, and clay was now a respectable material suitable for objects destined for the highest purposes.

2

The Han World of Figurines

THE First Emperor's army was the prologue to one of the two great ages of Chinese figurines. During the Han Dynasty (206 BC–AD 220), literally hundreds of thousands of these models were made. A lucky concatenation of events created a situation in which the potential of clay as a medium could be fully developed. The Han Dynasty was a period of economic expansion, and flourishing trade and agriculture led to a marked rise in the standard of living. It was an era in which people in all walks of life were obsessed with the idea of attaining immortality. In the new philosophical theories of Han Confucianism, filial piety — obeying and respecting one's parents — was a cardinal virtue and qualification for high office, and the desire to give visible proof of this virtue led to extravagant competition in tomb building. The need to satisfy the spirit of the deceased and the belief that part of the soul continued its existence in a world similar to this led to the construction of tombs whose plan, decoration, and furnishings gave an almost complete picture of contemporary Han life.

The use of life-size clay figures seems never to have been repeated, but pits with figurines continued to be used to enlarge the underground picture of this world given by the tomb and its furnishings. The more spectacular of these pit burials contain armies of figurines: thousands of miniature warriors placed like their larger Qin predecessors in perfect battle formation. Excavations in the late 1960s at Yangjiawang, near Xianyang, Shaanxi, unearthed 583 mounted cavalrymen and nearly two thousand figures of soldiers and civilians, around 50 cm high (Fig. 2.1 and Plate 3). Like the Qin army, they were placed in pits and flanked the

2.1 Standing and kneeling men, polychromed earthenware, Western Han Dynasty, *c.* mid-second century BC, reportedly from Xianyang, Shaanxi Province. Height 42.5 and 31 cm. Courtesy of the National Gallery, Prague.

southern approach to two mid-second century tombs of high-ranking officials. In 1984, pits with miniature clay horses and several thousand foot soldiers, kneeling archers, and civil attendants were found at Shizishan, Xuzhou, south of a royal tomb from the same period (Fig. 2.2). Like their Qin predecessors, these figurines were all made from clay, with clothing and armour painted after firing. Their military formation and clothing reflect differences between the Qin and Western Han armies: kneeling archers wear shoulder length balaclava helmets, and many of the infantry carry defensive shields (Plates 4 and 5).

In March 1990, however, an extraordinary imperial complex with quite a different sort of figurine was discovered near Xi'an. Twenty-four pits filled with figures were found in the area south of Yangling, tomb of the Western Han

11

2.2 Cavalry figures discovered in 1984 from a royal tomb at Shizishan, Xuzhou, Jiangsu, second century BC.

emperor, Jingdi (d. 141 BC), and his wife, the empress Wang. According to preliminary research, these pits cover 96,000 square metres, an area many times that of the First Emperor's army and roughly equivalent to twelve full-size football fields. Estimates of the number of figurines range from a minimum of 40,000 to as high as 100,000. Unlike the previous finds, these figures are unclothed and without arms (Plate 6). Remnants indicate that they would have been clothed in silk or hemp; flexible wood or bamboo arms, now perished, would have made it easier to clothe the figures than arms of clay. The features are finely modelled; bodies are painted flesh-colour, while the hair, eyes, and eyebrows are black. The idea of creating clothed figures with flexible arms seems to have come from the earlier wooden figurine tradition: the tomb of the Marquis of Dai, at Mawangdui, Changsha, Hunan, from about 168 BC contained both painted wooden figures and wooden figures fully clothed in embroidered silk (Plate 7). The new finds at Yangling include a wide variety of subjects. As well as military figures with miniature bronze and iron weapons, there are civilians with farm animals, horses and carts, and agricultural implements, all perfect replicas of real life objects one-third life-size.

Other 'outside burials' — the practice of burying goods in pits outside the tomb to complete an underground picture — were less military. A brick lined trench surrounding a minor tomb near Xianyang contained model hunters in the north-west corner; the inner wall on the south was guarded by painted pottery warriors with traces of straw sandals facing outwards over tightly packed rows of domestic animals — oxen to the west, goats in the centre, and pigs to the east. A pit near a royal tomb near Xi'an contained unclothed clay figures without arms, including twenty-four cavalry officials as well as eight male and female servants, the former accompanying bulls, the latter cows. Remains of the workshop in which these were made have been found in the north-east corner of the Han capital city, Chang'an.

Towards the end of the second century BC, however, outside burials were gradually abandoned. The use of pits had been developed for reasons of space — to hold large quantities of grave goods that could not be fitted into the tomb. In the early years of the Han dynasty, a large army was still an important feature of the state which needed to subdue rebel kingdoms seeking to break away from the centre, and it was therefore natural to continue Qin Shihuangdi's example. Once the Han were firmly established, however, they returned to the classical theories of Confucius based on the idea that true power rests on the consent of the people, not in the sword. In such a climate, the creation of vast underground armies no longer seemed appropriate. At the same time, an increasing use of symbolism made size and numbers less important. Having accepted that a miniature clay warrior could perform as well as his life-size counterpart, the Han now went further, giving a single stone figure the power through its symbolic qualities to replace a multitude. Through an association with immortality — stone endured, endurance was a form of longevity and thus of avoiding death — stone, previously ignored in tomb arrangements, became a popular material in all aspects of tomb furnishing. The earliest dateable stone monuments placed above ground on a grave — those on the tomb of Huo Qubing in 117 BC — are almost contemporaneous with the last underground army.

At the same time, a change in burial practices whereby certain burial rites were now performed inside the tomb, and husbands and wives, previously buried separately, were placed in the same grave, led to an enlargement of the tomb which gave greater space for grave goods. The need

14

to reopen the tomb after the first burial led to the increasing popularity of brick and stone chambers which weathered better than wood. These factors led to a fundamental reappraisal of the overall tomb plan. The multiple tasks of the tomb — to protect the body, to provide the deceased (and hence his descendants) with status, and to satisfy his spirit by recreating the known world — remained the same, but these functions were now divided between surface and underground monuments and furnishings. By the first century AD, the system of erecting spirit roads above ground, of lining the main approach to a tomb with stone monuments and figures of men and animals, was established, and these stone figures came to replace the underground armies. Through their supernatural attributes and contacts with the spirit world, a pair of stone felines, for example, symbolizing the ruler and military strength, were able to offer the protection and status of an unlimited number of clay warriors.

The task of recreating everyday life below ground was now concentrated within the burial chambers. Tombs were increasingly built to resemble surface dwellings. Vertical pit tombs were gradually replaced by horizontal tombs, dug into the hillside or built with brick or stone chambers, reproducing the different areas of a manor house which were then furnished with figurines. A recently excavated second century BC tomb near Xuzhou, Jiangsu, shows how literally this task was taken. The tomb, dug into a low hill, consisted of two adjoining complexes on different levels. The upper with antechambers, rear chamber for the coffin, and two lavatories represented the residential part of a royal palace. The lower level with banqueting hall, kitchen, storerooms, woodshed, ice cellar, arsenal, and well replaced the official and functional areas in the palace. Although the tomb had been plundered, more than a thousand objects were found, including 422 painted clay figurines. In niches off the entrance passage, tightly packed officials, attendants with ceremonial swords and emblems of high office, and archers with quivers on their backs provided a guard of honour. Within the chambers, male and female attendants, standing and kneeling, waited on the deceased, whilst dancers and musicians performed in the banqueting hall. There were weapons in the arsenal, pottery and lacquer vessels in the storerooms, a miniature stove and pots in the kitchen, charcoal in the woodshed, and grain in the storeroom. The attendants wore different styles of clothing painted in a variety of colours, and some of the women's robes were embellished with rich cloud and mountain patterns and intricate borders adorned with pearls and other ornaments on their sleeves, collars, and hems.

Chinese attitudes towards *mingqi* or objects made specifically for tomb use, have always been ambivalent. It was important to provide the dead with luxuries, but the presence of valuables in a tomb was an irresistible magnet for grave robbers. In the fifth century BC, Confucius posed the problem like this:

To treat the dead as dead would be inhuman; one cannot do that. To treat the dead as living would not be wise; one cannot do that either. Therefore one should furnish them with bamboo utensils which are unsuitable for use; ceramic vessels, but not washed; things of wood, but unfinished; zithers with strings tightened but not tuned; mouth organs with complete sets of pipes but unharmonic. . . . All these are called spirit utensils (*mingqi*) because we use them to serve the ancestral spirits.

In this context, clay was the perfect medium, having little intrinsic or reusable value. Through a combination of realism and symbolism, the Han tomb and its furnishings

2.3 Standing lady, Western Han, painted earthenware with slate grey wash. Height 47 cm. Courtesy of the Trustees of the Victoria and Albert Museum.

provided a picture of contemporary life and thought which has never been surpassed. Although never made explicit, there seems to have been a tacit division of subjects between stone and clay. Broadly speaking, the official aspects of life and spirit creatures were represented in stone, the activities and objects of daily life in clay. In two-dimensional stone and brick tomb reliefs, however, this distinction does not apply. These reliefs cover all subjects, dividing them spacially with spirit world and official scenes placed above scenes of everyday life.

Throughout the Han period generally, the most popular subject for figurines were people. At first predominantly military, these came to include civil officials, male and female attendants and servants, ladies of fashion, cooks, peasants, millers, and entertainers of every sort (Fig. 2.3).

As the field widened, two styles became clear. Whereas officials and attendants found in Western Han pits are usually modelled with the same meticulous realism as the clay warriors, other tomb figures follow the earlier Warring States tradition, emphasizing movement and form rather than detail (Fig. 2.4). The bodies are often out of proportion, little more than rough cylinders, but the figure as a whole catches the essential nature and occupation of the subject. For the modellers, the aim was to catch the spirit rather than the exact appearance. They are concerned with what later came to be known as *qiyun shengdong* (life movement through spirit consonance), the most important of the Six Principles evolved by the fifth century art critic, Xie He. Even today, modern sculpture is still judged by this criterion: a carving is good if it is alive; the posture

2.4 Dancing lady, second–first century BC. Height 45 cm. Shaanxi Provincial Museum, Xi'an.

and outline must convey the inner nature of that which is portrayed.

The figures give a clear picture of the various costumes of the time. Officials and male and female attendants of higher rank wear long cross-over robes of silk or gauze with full sleeves long enough to cover their hands when respectfully clasped in front; the robe is closed at one side and held in place by a belt or sash; contrasting colour borders are shown on the sleeves, collars, and hems. Layers of warm undergarments of different colours are worn under this, showing at the neck. The square toed shoes are exact replicas of a pair of hempen shoes found in a tomb dating from 167 BC. Female attendants stand or kneel with modest expressions; some ladies of fashion, with elaborate hairstyles and sometimes a mirror in their hands, sit crosslegged. As well as the long cross-over robe, women could also wear long-sleeved blouses over ground length skirts, often widely flared, and their hair, parted in the middle, is fastened in a loose bun at the back of the neck. Lower class servants and peasants wear knee- or three-quarter-length coats with narrow sleeves over trousers, their hair tucked into a small cap (Fig. 2.5). Whereas workers are often portrayed in action, complete with their implements — a man holding a spade or carrying a sack of grain — those of higher rank usually stand or kneel, holes in their hands indicating that they once held weapons or emblems of office. Some Western Han figurines were modelled with only one arm, a hole in the other shoulder indicating where the other arm, presumably of another material, would have been inserted.

The Han world was an agricultural one. In the early period, cavalry horses outnumber all other animals, but the horse was a status symbol valued for its use in hunting and war, never a farm animal, and as the emphasis on military display dwindled so did the number of horse

2.5 Man with a spade, first–second century AD, stone. Height 66 cm. Sichuan Provincial Museum, Chengdu, Sichuan.

figurines. Primarily associated with rank and ceremony, horses appear more frequently on stone reliefs than among tomb models. From the first century BC onwards, the 'celestial horses' from Ferghana in central Asia, prized for their speed and vitality, were imported for military and imperial use, replacing the local chunky Mongolian types. This change is reflected in horse figurines. Clay horses were now designed to convey a sense of power, and these Han horses are spirited creatures with lifted heads and perked ears. Slits in the neck and back indicate that real hair was inserted for manes and tails.

Early Han tombs were stocked with storage containers filled with meat, grain, and wine, but these were gradually replaced by models of food-giving animals and grain-processing machinery. Domestic animals such as pigs, goats

and sheep, oxen, and every kind of poultry are modelled with or without young, often in an agricultural setting. Two men with straw paniers lean over the wall of a pen to feed goats; pigs suckle their young in pigsties complete with a latrine above one corner, reflecting the practice of the time (Plate 8). There are farmhouses, sheds, granaries up whose steps peasants toil with their sacks of grain, and a variety of agricultural machines for husking and grinding (Fig. 2.6 and Plate 9). One of the more remarkable finds from the late Western Han is a model winnowing machine with a man turning the rotary fan which separates the grain from the husk; beside him another man works a tilt hammer to grind the corn. Wells are common, and their decorated superstructures often reflect the belief that a well, in which the world of the living was mirrored, provided an entrance to the world of the dead.

Tombs from rice-growing areas contain models of country scenes such as rice paddies with dykes and sluices, fish ponds with dams, and reservoirs filled with carp, frogs, snails, lotus, water calthrop, soft-shelled turtles, ducks and geese bred for food. Models of water conservancy systems found in four tombs in Shaanxi province are similar to many dams and canals in the region which have been in use since the Han dynasty.

Dogs were usually placed singly; often harnessed, they are clearly watch-dogs and sit inside the doorway or courtyard of large houses (Fig. 2.7). A few animals, such as bears offering protection against thieves, were included for symbolic purposes, and there are a few fabulous creatures, such as rhinoceros-like beasts with long single horns, included as a form of protection. By and large, however, Han animal figurines are confined to domesticated species designed to provide the deceased with food.

The Han period was an age of expansion and building

2.6 Han Dynasty granary, earthenware with lead glaze. Two guards sit outside while peasants carry sacks of grain up the staircase. The cocks on the roof are omens of good fortune. Height 95.2 cm. Courtesy of the Glasgow Museums: The Burrell Collection.

2.7 Watch-dog with harness, Western Han Dynasty, second–first century BC, earthenware with lead glaze. Height 33 cm. Courtesy of the Glasgow Museums: The Burrell Collection.

activity. Tomb models provide valuable evidence of contemporary architectural styles and building methods with regional variations, from simple single-story farmhouses or houses on stilts in the less-developed watery southern regions, to elaborate multi-story manor houses with watchtowers in the metropolitan areas around the capitals of Chang'an (modern Xi'an) and Luoyang (Fig. 2.8). In larger complexes the parts were sometimes removable as in a doll's house. In a six piece model from a tomb near Zhengzhou in central China, the separate buildings form a typical courtyard complex: there is a pig in the sty, a hen and three chicks on the granary roof, three small bronze vessels (a cauldron, bowl, and rice steamer) on the stove in the kitchen, and a dog on guard at the entrance. The model was carefully arranged in the outer chamber of the tomb with the gate and main hall on a true north-south axis.

2.8 Han Dynasty house with a tower, Hubei, red earthenware with green glaze. Height 72 cm. (From *Quanguo Chutu Wenwu Zhenpin Chuan* 1976–84, Beijing 1985.)

Even more elaborate are the towers found in tombs in central China from the first and second centuries AD. Contemporary literature abounds with references to tall towers used as granaries, pleasure pavilions, and watch-towers for defense (Fig. 2.9). Towers symbolized wealth and rank; through their height they provided a possible means of contacting the immortals, and by the second century towers were placed in nearly all rich tombs. The three most common types are tall grey unglazed towers with painted architectural features, green glazed towers in basins simu-lating water pavilions, and very tall green glazed towers, up to 1.5 metres high, peopled with small moulded figures of human beings and birds or animals. Towards the end of the period, towers become increasingly fanciful. A multi-story tower in a bowl representing a water pavilion from an Eastern Han tomb in Henan seems to reflect the ideal-ized life a wealthy landowner. Protected by an archer with bow drawn, the tomb occupant kneels in the pavilion, two servants with brooms standing nearby. On the rim of the

2.9 Water pavilion, first–second century AD. While the tomb occupant enjoys the view, a lookout keeps guard from the upper story. The roof beams and brackets are formed like mythical animals to attract immortals. Height 86.5 cm. Courtesy of the Trustees of the British Museum.

bowl, symbolizing dry land, are two men talking, a horse and rider, a man with his feet in the pool, three deer, three geese, two ducks, and a pair of birds on a perch. In the water below the bridge connecting the pavilion with land is a small boat complete with paddle, pole, and holes for a mast and rudder, whilst turtles, fish, and waterfowl swim around.

These architectural figurines are a Han speciality. The Ming produced fine architectural models but confined themselves to the multi-courtyard mansions of the wealthy. Model buildings from other periods are few and with little detail.

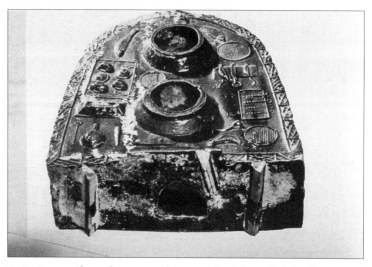

2.10 Stove with two burners, cooking utensils, fish, and kebabs, first–second century AD, dark green glazed earthenware. Height 9 cm. Courtesy of the Royal Ontario Museum, Toronto, Canada.

Finally, in Han tombs the deceased were feasted and entertained. There are models of stoves complete with fish, snails, or kebabs on skewers, cooking pots, and implements (Fig. 2.10). Tables are laid with bowls of food, wine cups and ladles, with chopsticks and knives on the floor nearby. Gentlemen play *liubo* on a patterned board with draughtsmen and bamboo markers, a game believed to be a popular pastime of the immortals. The multi-cultural character of Han entertainment shows in the wide variety of musical instruments, which include not only traditional Chinese

1. Painted wooden figure, Xinyang, Henan, eighth–fourth century BC.
Height 64 cm.

2. Kneeling figure found near tomb of Qin Shihuangdi, Shaanxi. Height 67 cm.

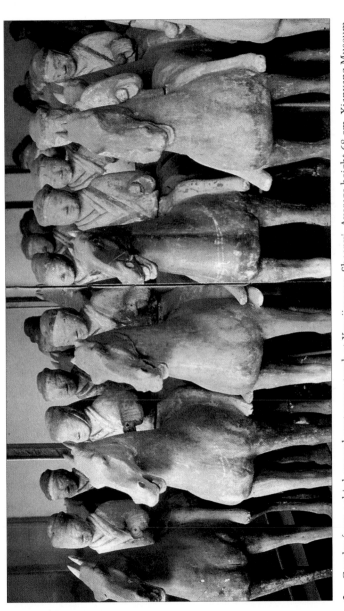

3. Cavalry from a third–second century BC tomb at Yangjiawang, Shaanxi. Average height 68 cm. Xianyang Museum, Shaanxi.

4. Male attendant with sword, first–second century AD, Sichuan. Height 99 cm.

5. Kneeling archer with hood, second century BC, Xianyang, Shaanxi. Height 25 cm.

6. Unclothed warriors found in 1990 near tomb of Han Jingdi, (d. 141 BC), Xi'an, Shaanxi. Height *c.* 60 cm.

7. Wooden figure with silk clothes, Mawangdui, Changsha, Hunan, 168 BC. Height 79 cm.

8. Pigsty with latrine, first–second century AD, lead-glazed red earthenware. Height 16 cm. Courtesy of the Royal Ontario Museum, Toronto, Canada.

9. Granary with trip-hammer, Eastern Han. Courtesy of the Royal Ontario Museum, Toronto, Canada.

10. Acrobats and jugglers performing to music while the audience watch from the sides, second–first century BC. Height 22 cm. Jinan City Museum, Shandong.

11. Story-teller with a drum, first–second century AD, earthenware. Height 66 cm. Sichuan Provincial Museum, Chengdu.

12. Story-teller with clappers, stone, first–second century AD, Shandong.
Height 31 cm.

13. Story-teller with drum, first–second century AD, earthenware. Height 55 cm. Sichuan Provincial Museum, Chengdu.

14. Animal guardian, reportedly from a tomb in Luoyang dated 525, painted earthenware. Height 24.6 cm. Courtesy of the Royal Ontario Museum, Toronto, Canada. The George Crofts Collection.

15. Animal guardians, guardian kings, and civil officials, reportedly from tomb of general Liu Tingxun, (d. 728), near Luoyang, earthenware with glaze and pigments. Height 1.7 m, 1.03 m, and 0.96 m. Courtesy of the Trustees of the British Museum.

16. Civil official, early seventh century. Shaanxi Provincial Museum, Xi'an.

17. Camel with travelling musicians, seventh–eighth century, glazed earthenware. Height 66 cm. Shaanxi Provincial Museum, Xi'an.

18. Mounted foreigner, unglazed earthenware, seventh century. Height 37 cm. Shaanxi Provincial Museum, Xi'an.

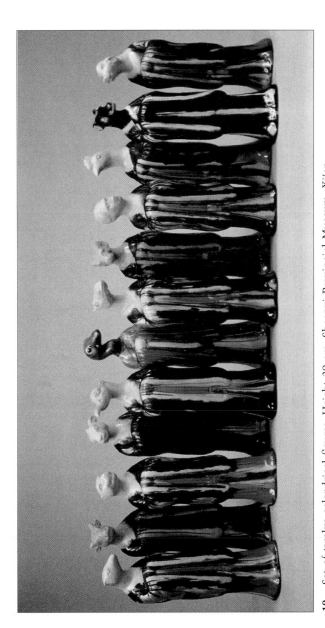

19. Set of twelve calendrical figures. Height 30 cm. Shaanxi Provincial Museum, Xi'an.

20. Three-courtyard house, reportedly from Henan. The number of courtyards in a house depended on the rank of the deceased; only those of princely rank might have three-courtyard houses. Sixteenth century, earthenware with glaze and pigments. Height 68.5 cm. Courtesy of the Royal Ontario Museum, Toronto, Canada.

2.11 Man playing a stringed instrument, first–second century AD. Height 36 cm. Guizhou Provincial Museum.

types such as the horizontal zither (Fig. 2.11) but also disc-drums and reed trumpets introduced by non-Han peoples from the north-west. It was common to combine dance, music, and acrobatics. While graceful female dancers with long sleeves revolved to the music of seated musicians, acrobats and jugglers performed feats which can still be seen today. At the Western Han court the 'Yellow Gate' bureau was set up to provide the emperor with painters, scholars, astrologers, jugglers, wrestlers, and fire-eaters who could be summoned at any time of night or day. The aristocracy kept their own troupes of entertainers, and there were travelling bands wandering from one province to the next. A complete set of figures on a base from a tomb in Shandong shows guests watching a mixed performance by dancers, musicians, and acrobats backed up by a row of drummers and percussion players (Plate 10).

Some of the most lively figurines of all are jesters or story-tellers from tombs in Sichuan. Story-tellers were popular at court, and these figures, often dwarfs grimacing and exaggerating their deformities for comic relief like court jesters in the European Middle Ages, were found in manor houses and small tea rooms throughout China (Plate 11). With their characteristic half-chanting half-singing (*shuochang*), they accompanied their narrative recitals with musical clappers or drums, often dancing at the same time (Plates 12 and 13). Such story-tellers can still be found in the countryside today.

This world of figurines is a cheerful place. It is a mirror of the world as it ought to be. A man could not change his status in the next world through tomb furnishings, but he could improve his lot. Figurines provided him with what he would like to have had rather than what he actually possessed. In keeping with the idea that images and models have the power to influence their surroundings, there were certain unwritten restrictions on their use. As mentioned earlier, there was a clear demarcation between subjects suitable for stone and those modelled in clay. The world of imperial officialdom belonged to stone; there are no clay figurines of the fine hunters and chariots portrayed in stone tomb reliefs, no representations of imperial occasions or insignia, and little reference to the land of the immortals. The conventions of spacial realism were observed. In stone murals, for example, winged spirits were always shown above scenes from this world. Clay figurines were placed on the tomb floor, on the ground, belonging quite literally to the daily life of ordinary, not elevated, people. They reflect contemporary Confucian attitudes and morality. Apart from the entertainers, expressions and gestures are moderated: ladies smile rather than laugh, and unless dancing, adopt demure postures.

As in a Victorian doll's house, there are no scenes of immorality or explicit sex, no villains and punishments. Unlike the almost contemporary bronzes of the non-Han Dian people in south-west China, there are no representations of death, illness, mutilation, or violence. More surprising, there are no mothers with babies and seldom any children. The rare figures of children that do appear are thought to represent the soul of the deceased.

These Han figurines were not created as works of art; they were made to answer the needs of a particular belief about life after death and the spirit world. Nevertheless, they are the embodiment of the indigenous Chinese talent for sculpture. The robust approach and treatment of the material, the concentration on catching the life and vitality of the subject rather than striving for a meticulous portrait, and the manner in which forms are modelled to catch movement reflect an extraordinary mastery of the art of three-dimensional representation. The elegant Tang figurines of the seventh and eighth centuries may come closer to Western ideas of art, but never again was the atmosphere and reality of daily life so successfully caught as under the Han.

Techniques and Production Methods

The production of Han tomb figurines was based on the techniques evolved for mass-producing moulds for ritual vessels in the Bronze Age. Like the life-size Qin Dynasty warriors, the miniature figures in the Han underground armies were hollow and made from moulds — front and back for human beings, left and right for animals — which were then joined. The seams were smoothed and finer details incised in the wet clay before firing. Clothing details were incised or painted with coloured pigments on white slip (a thin paste of clay mixed with water). With non-military figures and the development of a distinctive Han style, the need for detailed modelling and incising lessened, but the basic method of combining moulded and modelled pieces remained, providing a means for creating mass-produced figures with realistically individualized features. Certain house models appear to have been made entirely by hand, details in the house such as tile ridges, lattice windows, or openwork doors being cut with a sharp knife while the clay was still wet. As the dynasty progressed, however, rising standards of living led to increasing demands for figurines, and as always, the popularization of a status symbol brought a decline in quality. Increasing use of mass production in the Eastern Han led to a certain stereotyping and even stiffness.

Most Western Han figures are unglazed, their colour varying with the colour of the original clay. Grey figurines were the result of the 'reduction' method, whereby the atmosphere within the kiln was starved of oxygen, producing a tough grey material which was then covered with white slip and painted with coloured pigments, usually red and black.

Although a few glazed pieces have been found from the first century BC, widespread use of glazing belongs to the later Eastern Han period and was concentrated in the central regions of Shaanxi, Henan, and southern Shanxi. Lead, which had been used in bronze alloys to lower the melting point of the base material, gave a glaze which was both decorative and protective. The earliest glazes were brown or yellowy-brown when iron was added to the glaze mixture. Soon, however, copper-oxide came to be used to produce a brilliant green colour; when exposed to air and water this degrades into the silvery-grey iridescence seen on so many Eastern Han figurines. The use of glazes affected attitudes towards colouring. Whereas the colours on unglazed figurines kept close to the original, there was little attempt to match reality with the colour of a glaze, and a dog, for example, could as well be green as brown.

Production was concentrated in officially sponsored factories around the capitals, Chang'an and Luoyang, and in provincial centres in Gansu, Sichuan, Hunan, Shandong, Guangdong, and Jiangsu. Like almost all aspects of tomb planning, figurines were usually ordered before death, the choice reflecting the rank and interests of the individual.

During the following centuries, improvements in pottery firing and glazing led to finer wares and a wider variety of colours, but the basic method of production and the combination of mould-using mass production with individually modelled pieces continued throughout the history of tomb figurines.

3

The Northern Dynasties:
A Period of Transition

OVER three hundred years were to pass before tomb figurines regained the importance they had enjoyed under the Han. The fall of the Han Empire was followed by a period of disunion with thirty different dynasties contesting power in 360 years. Reacting against Han funeral extravagance, the early post-Han rulers forbade expensive burials, banning the use of tomb statuary above ground and the burial of jade, gold, and treasure in the tomb. From the early fifth until the late sixth centuries, China was divided into two: the Northern Dynasties under non-Han nomadic peoples from the steppes, and the Southern Dynasties, under traditional Chinese rulers, whose capital was centred on Nanjing. This division led to a divergence in funerary habits and tomb arrangements. In the north, Buddhism became the official religion, and there was a certain transference of ancestor worship from the tomb to the new Buddhist shrines at Yungang and Longmen. Few large tombs were built, and only a handful of stone spirit road statues have been found. In the south, ancestor worship at the tomb remained at the heart of the imperial system, and imperial and royal tombs were once again provided with stone guardians.

   Throughout China, small numbers of tomb figurines have been found in tombs from the Western Jin Dynasty (265–316). During the period of division, however, the earliest examples of sizeable groups of figurines are from the tombs of Chinese officials working for the alien rulers in the north. True to tradition, they reflected contemporary life above ground and, like the northerners themselves, introduced

3.1 Horseman with balaclava hood and high collared long coat, Northern Wei Dynasty, *c.* fifth century. Height 21 cm. Courtesy of the Trustees of the British Museum.

new elements into the Chinese tomb world. The northerners were nomads, horsemen from the steppes with contacts across the deserts to central Asia and further west (Fig. 3.1). Their figurines reflect a world of travel with processions, beasts of burden such as packhorses and camels, mounted warriors, and hunters. The tomb no longer resembled a dwelling but consisted of a long shaft leading to a single hall or occasionally a hall with an antechamber. Figurines forming guards of honour were placed in niches off the shaft tunnel; others stood in the antechamber or the outer end of the burial hall.

A fourth century tomb at Caochangpo, near Xi'an, contained a model procession centred on an ox cart surrounded by infantry and cavalry wearing belted tunics and baggy trousers; before and behind were mounted musicians with drums and pipes. In the antechamber of the same tomb were twenty male and female servants and seated female musicians. In another fourth century tomb in Inner Mongolia, many of the figures were non-Han, representing the Xianbei, the nomadic forerunners of the Northern Wei who controlled northern China from 386 to 534. The military wear helmets and sets of breast and back plates over a protective skirt; civilians wear a distinctive round hat or balaclava type hood and high collared coats draped over their shoulders, the sleeves hanging empty. The tomb of a Chinese-born official, Sima Jinlong, who worked for the Northern Wei and was buried near their capital Datong in 484, contained 367 small figures of mounted warriors and male and female attendants, all with non-Han features and in northern dress. There were also 33 ceramic animals, including two packhorses with grain on their backs and three camels. Camels were an important addition to the figurine world. The two-humped Bactrian camel was the main means of transport across the Tarim and Gobi deserts on the Silk Roads to the west.

In 494 the Northern Wei moved their capital south to Luoyang and followed a deliberate policy of adopting Chinese traditions. From this time on, the use of figurines increased and the scope of subjects widened. The continual influx of foreigners along the trade routes from the west is reflected in models of central Asians or Indians with curly hair and beards. Figures of civilians and of women begin to outnumber those of warriors. At the same time, the role of figurines was expanded to serve protective as well as representational purposes. Although a few examples of fabulous beasts with long horns have been found in late Han tombs, the prime task of Han clay figurines was, as we have seen, to reproduce real objects; supernatural protection was provided by stone carvings or decoration on costly objects such as jade or lacquer. The Southern Dynasties maintained this distinction: clay figurines were based on reality; contact with or help from the supernatural world was obtained from gigantic stone fabulous beasts above ground and stone

3.2 Animal guardian, Northern Wei Dynasty, *c.* fifth century, painted earthenware. Height 25 cm. Courtesy of the Trustees of the Victoria and Albert Museum.

warriors carved in relief on the tomb entrance below ground. The northerners, however, ignored these distinctions of media, and in an attempt to match the protection afforded in southern tombs, introduced clay animal and warrior guardians as counterparts to the southern stone figures.

One of the earliest examples of northern animal guardians is from Sima Jinlong's tomb. Crouched on its haunches, it has a human head, truncated horn, and five holes in the back for hair. Made of unglazed clay, the face is painted white, and white lines on the body indicate scales. By the early sixth century, the use of a pair of sitting animal guardians was common in wealthy tombs. These creatures, known as *zhenmushou*, were hybrids: one had a human face, the other a beast's head; both had feline bodies with long tails and three upright spikes in the back (Fig. 3.2 and Plate 14). Like the mythical stone creatures in Southern Dynasty spirit roads above ground, their fabulous appearance indicated the power to draw on help from the other world to protect the deceased.

At the same time, a pair of Northern Dynasties clay warrior guardians matched the armed figures carved on stone or brick portals to Southern Dynasties tomb chambers. In Chinese sculpture, status is indicated by height, and these guardians flanking the entrance to the tomb chamber were markedly taller than any other figurines. Appearing first in the early sixth century, they were common in upper class tombs in the north from the mid-sixth century onwards. Early examples, often with one hand resting on a shield placed in front of the left leg, are realistic representations of a group of specially trained warriors appointed as guards for senior officials.

Stylistically, figurines from the Northern Dynasties period fall into three phases. The early figurines from the tombs of Chinese officials in the north lack the Han sense of movement but retain their close link with reality. The warriors are based on real people, on thickset men from the steppes; their mounts are sturdy Mongolian ponies with heavy legs.

After the move south to Luoyang, there is a distinct change in style, and the influence of Buddhist rock carvings becomes apparent. Accustomed to Buddhist cliff sculpture, which was mainly intended to be seen from the front, modellers began to adopt a purely frontal approach. Details were concentrated on the front, carved delicately in planes of low relief, giving an almost two-dimensional effect. Unlike the Han figurines made from two hollow moulds, some Northern Wei figurines were solid, made in a single shallow mould. Sometimes the head formed part of the mould; more often it was made separately and fitted with a socket so that it could be turned. This single mould or 'cookie-cutter' method was used particularly for civil officials and male and female attendants who stand, sit, or kneel in strongly frontal and symmetrical positions, usually static with legs carved together. The arrangement of their clothing and facial expressions mirror the iconography of contemporary Buddhist stone figures. The folds and pleats of clothing reflect Buddhist sculptural styles, and for the first time in figurines a distinction is made between the body and its clothing, enabling the viewer to perceive the human form beneath its dress. The robust realism of Han facial expressions is replaced by a subtle characterization of emotion and serenity. Female attendants with elegantly arched eyebrows, small noses, and oval or square faces on long necks smile enigmatically. Male officials, sometimes nearly a metre tall and often with fat Buddha-like bellies, exude a Buddha-like repose. Proportions also changed: human figures were given small heads, long cylindrical necks, and attenuated bodies. A similar attenuation in the animals produced elegant but stylized creatures with little connection to reality. Horses from the late fifth and early sixth centuries have small pointed heads on long arched necks thickening at the base to join a broad chest. Lacking all musculature, these richly caparisoned beasts had such spindly legs that they had to be supported by a base.

The third and final phase of the Northern Dynasties figurine style appears during the mid-sixth to early seventh centuries when sculptors reverted to the more traditional Chinese styles and figures again became rounded with realistic proportions and details modelled in full relief. At the same time, technical developments in ceramics led to the production of high-fired wares of a finer quality than any yet seen. Most northern figurines are unglazed, made from iron-rich clay covered with a white slip and then painted in many colours. From the late fifth century onwards, however, a renewed interest in glazing led first to the revival of green and brown lead glazes and then to experimentation with polychrome glazes such as green under yellow.

4

Sui and Tang: The Second Flowering

THE second great age of tomb figurines lasted from the re-unification of the empire under the Sui (581–618) through the first half of the Tang Dynasty (618–907). Unity, military security, and stable, efficient administration brought peace and unprecedented prosperity. This was a great empire, filled with energy and self-confidence, and in the early seventh century the capital, Chang'an, was the largest and most cosmopolitan city in the world. Tang figurines reflect the wealth, vitality, and openness of this great empire.

Once again the tomb became a powerful political instrument used to reinforce the might of the central government, and the Tang imperial tombs are among the most splendid in the world. The second emperor, Li Shimin, set the pattern by choosing a mountain 1,200 metres high for his tumulus. Above ground, his tomb was based on the capital city with high walls and gate towers and over three hundred halls and dwellings within an outer wall 60 kilometres long. Below ground, the Tang developed the Northern Dynasties tomb pattern with a long ramp leading to two vaulted chambers, the first resembling a courtyard and the inner one housing the coffin. The ramp was ventilated by vertical shafts and between these, just above floor level, were niches for clay guards. Murals on the ramp and chamber walls recreated the elegance and luxury of a highly sophisticated society. This was no longer a house below ground; the underground tomb became a landscape with murals depicting the palaces, gardens, and open country-side in which the nobles passed their lives.

Tomb figurines occupied a more important role than before. The inclusion of the Northern Dynasties protective guardians led to the development of a figurine guard of honour — often a set of ten — whose functions were similar to those of the stone statues lining the spirit road above ground: providing protection, links with the spirit world, and status. The majority of figurines were still representational substitutes for real life persons or objects, but the choice of subjects was influenced more by the desire to give the deceased status than to recreate daily life. In place of the Han world of settled farmers with its emphasis on agricultural and domestic pursuits, figurines now reflected a sophisticated world of wealthy nobles from the north, accustomed to foreign contacts and travel and enjoying all the luxuries of aristocratic life in the capital city. As under the Northern Dynasties, the underlying theme was mobility.

There is a gradual evolution in choice of subjects. Until 683, the main emphasis is on tomb guardians and processions in which the chief personage is carried in an ox cart escorted by military figures and entertainers (Fig. 4.1). Over the next thirty years, horses gradually replace ox carts as the focus for groups of ceremonial attendants, and not until the mid-Tang, from 711 to 779, does interest shift from travelling to the more informal aspects of domestic life depicted in youthful attendants, garden villas, and rockeries. In general, the most popular subjects are human beings: wealthy officials or nobles with their myriad attendants. Horses come next and then camels which, with the foreign entertainers and merchants, reflect foreign contacts and interest in the outside world.

In keeping with their increased importance, figurines were prominently displayed during the funeral procession. For an important dignitary, the route to the cemetery was lined with a guard of honour, and the coffin, pulled with long ropes by chanting bearers and escorted by mourning family and officials, was preceded and followed by musicians

4.1 Ox cart which would have formed the centre-piece of a cortège of soldiers and musicians, Tang Dynasty, seventh century. Height 25 cm. Courtesy of the Ashmolean Museum, Oxford

and other entertainers. The figurines were placed upright in open carts, their size and number indicating the rank of the deceased. On arrival, they were lined up outside the tomb while the coffin was carried into the chamber; then they were placed in their correct positions undergound. Thanks to recent excavations of unplundered tombs, it is becoming clearer where the different kinds of figures stood. In a recently excavated Sui tomb near Taiyuan, 328 figurines were found mostly in the tomb chamber. However, the finest figures, the guard of honour of military and animal guardians, camels, ceremonial horses, and mounted musicians, stood outside it and faced up the entrance ramp. In early Tang royal tombs, many hundreds of tightly packed earthenware mounted hunters and warriors were found in small niches in the tomb ramp. Within the tomb, the more important figures stood close to the coffin or tomb tablet (a stone slab inscribed with a eulogy to the deceased) or in strategic positions flanking the entrance.

An official bureau, the *Zhen Guan Shu*, supervised the production and use of burial goods. Figurines were used in sets, the number, size, and content of the set depending on the rank of the deceased. The most important set, a special guard of honour consisting of figures distinguished by their size and the excellence of their material, colour, and glazing, was reserved for the highest ranks. The exact composition of this guard of honour varied. Usually comprised of ten figures, it always included a pair each of animal guardians, military guards, and civil officials (Plate 15). To these were added one or two horses with grooms and/or a camel with a groom. Ten was a recurring figure in Tang court etiquette. Above ground, the stone guard of honour in the spirit road consisted of ten ceremonial horses and grooms and ten pairs of officials, corresponding to the imperial guard in real life.

Within the set, those with a protective role had pride of place. The tallest, and therefore most important, were the military guardians. By the mid-seventh century, the widespread popularity of the four Buddhist Guardian Kings, or *lopakalas*, led to these figures being assimilated with the four Chinese 'Heavenly Kings', legendary guardians of the four directions, and adopted as models for tomb guardians. Unlike their predecessors, the burly unglazed warriors of the Northern Wei, these were no longer based on real human beings but are clearly supernatural. Their ferocious expressions and menacing gestures are borrowed from their Buddhist counterparts, and their hair is frequently pulled up into a distinctive knot in the fashion of Buddhist deities. Up to 1.5 metres tall, they trample on evil in the form of a small demon, or they stand on an ox or cow, symbolizing that the king is the guardian of the south.

The animal guardians enjoyed a similar promotion. The rather small unglazed human- and lion-faced hybrids of the late sixth century became large glazed creatures with the body of a crouching lion, spiky wings on the forequarters, cloven hoofs, and a long curled tail. Three flamelike spikes and sometimes a horn of twisted hair rise from the back and head, and a two-pronged lance fixed behind the head emphasizes their protective role. By the mid-eighth century, they frequently stand upright, the hind leg crushing a small creature, and the foreleg raised about to strike in imitation of the posture of the Guardian Kings. Later still, these hybrids are reduced to small symbolic figures draped round the necks of a pair of human warriors. The importance of these mythical creatures reflects the rearrangement of protective roles in the tomb which had begun under the Northern Dynasties. Tang stone statuary above ground is primarily concerned with reality, and the protective and supernatural role of the earlier Han and Southern Dynasties surface statues of winged beasts is concentrated in these clay figures below ground.

This interplay between surface statuary and figurines is carried further by the introduction of a pair of officials (Plate 16). Above ground, stone officials had always represented both the civil and military branches of government. This was now repeated below ground, with an impressive civil official in court dress placed opposite his military counterpart. Less important than the Guardian Kings, these officials are usually somewhat shorter.

The stylized iconography of these three pairs of figures gives them an air of theatrical unreality. With the remaining components of the guard of honour, the horses and camels, however, the group comes to life.

For the Tang, the horse was a symbol of status and power.

Skilful use of cavalry had given them control of the empire, and the horse was recognized as the era's most valuable military asset. Riding was an aristocratic privilege reserved for the imperial and upper classes. Merchants and artisans were forbidden to ride, although they might use horses for transport. When the dynasty started in 618, there were only 5,000 horses grazing in the grasslands of Gansu province; by the middle of the century, there were 706,000. All the Tang love of horses comes alive in the horse figurines (Fig. 4.2). The largest examples, up to 1 metre tall, are beautifully modelled, perfectly proportioned animals with exquisite colouring and glaze. Saddled or unsaddled, they are shown in every imaginable position: grazing, galloping, heads tossing and hoofs pawing the ground, or at rest with head turned as if to scratch a hind leg. The modellers knew their subjects — these are real animals with strong musculature and flared nostrils. The privileged status of the horse led to a saddled horse replacing the ox cart as a means of transport for the deceased, and from the early eighth century onwards, they stand as a centre-piece in the cortège of soldiers and musicians. These ceremonial horses have their manes cut in 'three flowers' or tufts, an imperial privilege, and the tails are turned up and braided. Stationary horses almost always turn their heads to the left, a result of dressage since horses were mounted from that side.

As northerners, the Tang were at home on horseback, and both men and women rode for pleasure, for hunting, and for playing polo. Mounted hunters in the ramp niches of early royal tombs carry quivers, bows, and swords; some have falcons or hawks on their wrists, dogs on their laps or behind their backs, and game slung across their saddles. As in most wealthy societies with a leisured class, there was a lively interest in foreign innovations and

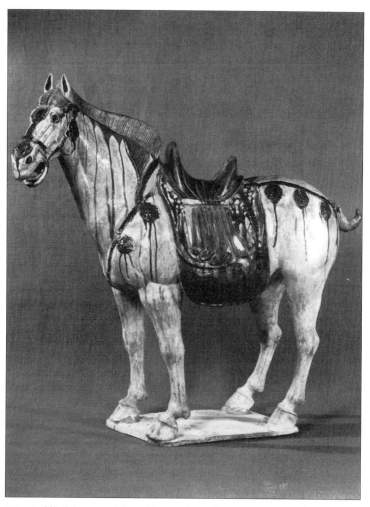

4.2 Saddled horse, pinky-white with amber mane and tail and green saddle, Tang Dynasty, 700–750, earthenware with coloured glazes. Height 74 cm. Courtesy of the Trustees of the Victoria and Albert Museum. Given by Mrs. Robert Solomon.

4.3 Equestrienne dancer, Tang Dynasty, *c.* 680–700. Height 42 cm. Courtesy of the National Gallery, Prague.

fashions. The popularity of polo, brought from Persia, is reflected in tomb murals and numerous figurines of male and female polo players shown at full gallop. The women often wear men's clothing for greater freedom of movement, their hair protected by a wimple (Fig. 4.3).

The same liveliness characterizes the camels. These, too, were a vital part of northern life, and the sculptors show complete familiarity with their subject. The figurines bring alive the flavour of caravans bearing goods and peoples from distant lands. Often one of a pair is laden, one unladen. Long boards strapped beneath the saddle bags were the supports for central Asian yurts used on the journey. The modelling is fine and detailed: fierce masks on the saddle bags repel evil spirits, and sometimes a camel carries an entire troupe of musicians, highly individualistic, bearded, large-nosed Westerners with their instruments (Plate 17).

As in real life, the camel drivers, stable boys, and grooms were foreigners, brought with their livestock from central Asia. Foreigners were a common sight in Chang'an, where the streets thronged with traders, entertainers, missionaries, and wealthy envoys arriving via the busy trade routes from the west or by sea from the east (Plate 18). Like the earlier Han figurines, these foreigners are portrayed with a startling sense of characterization. Based on acute observation of real life, the sculptor has concentrated on creating a living example of a particular type, catching the very essence of the subject — a merchant from western Asia bearing his wares in a sack over his shoulder, a felt-hatted Khotan, or one of the numerous dancing boys sent as tribute from states further west (Fig. 4.4). The expressions and features of such figures are so realistic that they are sometimes referred to as individual portraits, but this is not the case, as the same figures reappear in different tombs.

Among the Tang, there was a keen interest in fashion. Both men and women adopted foreign dress, hairstyles, hats, and accessories from Persia, Turfan, southern Asia, and Manchuria. Large tombs contained numerous female figurines — well-born ladies with their attendants, serving maids, and female dancers and musicians. Women's clothing ranged from daring *décolleté* to masculine coats with side slits and turned back lapels. This was a world in which women pleased. Beautiful girls were sent to the court as tribute, those from Kucha in central Asia being particularly famed for their music and grace. Young girls with slender waists and flowing sleeves dance seductively to the music of kneeling girls playing pipes, flutes, and zithers (Fig. 4.5). Smiling maidservants with 'butterfly eyebrows' bear dishes of fruit, trays with food, or bottles of wine; others bring fans, bags, shawls, or musical instruments (Fig. 4.6). Early female attendants are slender and willowy,

4.4 Dancing boy from Central Asia, Tang Dynasty, 680–750, painted earthenware with traces of gold leaf on the necklace. Height 27 cm. Courtesy of the Trustees of the Victoria and Albert Museum.

wearing low cut dresses and high waists with foreign style shawls draped round their thin shoulders. In the mid-eighth century, however, the example of the famous beauty, Yang Guifei, consort of the emperor, Xuanzong (d. 756), led to a preference for plump rounded figures, the so-called 'fat ladies' wearing loose flowing robes (Fig. 4.7).

Interest in fashion was not confined to clothing. Hairstyles vary from early examples of women with long hair down their backs to the later extravagantly coiffed ladies with bouffant hair and a loose chignon. Tang beauties were renowned for their use of scent, and some figures still bear traces of make-up: lips painted red, foreheads coloured yellow with powdered lead or arsenic, and painted Indian inspired flower-shaped beauty patterns between the eyebrows or on the cheeks.

The more domestic subjects of Han figurines appear in

4.5 Kneeling female musician with zither (*qia*), Tang Dynasty. Height 16.5 cm. Courtesy of the Trustees of the Victoria and Albert Museum. J. G. Maxwell Brownjohn Bequest.

4.6 Serving ladies holding a fish on a plate, a gourd, a musical instrument, a flask, and a censer, Tang Dynasty, late seventh century, painted earthenware with transparent glaze. Height *c.* 25 cm. Courtesy of the Glasgow Museums: The Burrell Collection.

Tang tombs but only in limited quantities, and their inferior status is reflected in their size and quality. Apart from the prestigious ox carts, the finest examples are of dogs, popular for hunting, in a wide variety of lifelike poses. There are also stoves, wells, trip hammer pestles, and domestic animals such as ducks, chickens, high-backed pigs suckling their young, and sheep. These figures are often unglazed and stand in secondary positions inside the tomb.

Finally, a new element was added: a set of twelve calendrical figures (Plate 19). Like the guardians, these were non-representational and created to perform specific supernatural functions. The figures, usually hybrids with human bodies and animal heads, represent the Chinese twelve year cycle in which each year is associated with an animal from the following series: rat, ox, tiger, rabbit, dragon, snake, horse, goat, monkey, chicken, dog, and pig. The earliest known reference to the twelve year cycle is in bamboo slips found in a third century BC tomb. The cycle was used in divination, at first for state purposes but later by individuals believing that character and fate were influenced by the year of birth. This form of fortune-telling became increasingly popular with the spread of Buddhism. A fifth century Buddhist text found near Dunhuang refers to the animal cycle, claiming that the animals, living in twelve caves previously occupied by Bodhisattvas, took turns to roam the world, helping those born under their sign.

Associated with the different compass points, these calendrical figures may have served to anchor the tomb in its correct place in the cosmos. The earliest known pictorial representations of the twelve are from a Northern Wei tomb of the Cui family at Linzi, Shandong, from 524, where they stand in niches with pointed arches like Buddhist haloes around the tomb chamber walls. Their supernatural or spirit-like powers are made clear in a tomb from 570 where they

4.7 'Fat lady', Tang Dynasty, *c.* 720–750, white painted earthenware. Height 46 cm. Courtesy of the Trustees of the Victoria and Albert Museum.

appear as cosmic deities in a relief round the lower edge of a ceiling vault decorated like the heavens. A Sui tomb from 610 contained two calendrical sets, both with seated figures, one with animal heads, the other of human beings with miniature animals standing on their shoulders and looking over the tops of their hats. In other sets from this period, seated human beings with tall hats hold the calendrical animals in their arms. Throughout the early Tang, they were shown standing, with human bodies and animal heads, but in the tenth century, they suffered a transformation similar to that of the Tang animal guardians, becoming officials in court dress, holding miniature animals like emblems of office.

After the mid-eighth century, the use of figurines declined. A powerful rebellion led by the military commander, An Lushan, fatally weakened the dynasty, and this was followed by the persecution of the Buddhists in the mid-ninth century in which statues made of valuable metals were melted down for reuse. As the economic situation deteriorated, attempts were made to curb tomb extravagance. A series of decrees in 742, 806, 841, 871, and 936 limited the use of burial objects, forbidding the burial of gold, silver, or bronze objects in tombs. The edict of 742 restricted the number of clay figures for those of the third rank and above to 90, for the fifth rank and above to 70, and for the ninth rank and above to 40. Later these quotas were reduced to 70, 40, and 20, whilst common people were allowed only 15 unglazed earthenware pieces. The same decrees set limits on the height of different categories and prescribed whether they might be glazed or unglazed or made of wood. These decrees were not always observed; there are numerous records of elaborate burials during the late Tang, and gold, silver, and fine brocade have been found in tombs from this period. The production of tomb figurines, however, fell sharply. In a definite change in tomb practices, there was an increasing use of wooden and metal objects or paper effigies which could be burned during the sacrificial rites.

Tang figurines reached their peak in the first half of the eighth century, and examples from this period combine all that is best in Chinese sculpture. Freed from the iconographical requirements of Buddhist sculpture with the frontality, symmetry, and immobility which had influenced Northern Wei figurines, the Tang combined Han concentration on the inner nature of the subject with their own distinctive genius for accurate reproduction. The result was fully rounded figures caught in the moment of action. The mounted hunter twisting in his saddle to shoot a bird, or the polo player leaning forward as his horse gallops after the ball, are living vignettes, conceived as a single unit with perfect balance between the parts. The proportions could be varied to produce the desired effect, but the result is always harmonious.

The Tang period is the high point in the history of Chinese tomb figurines. The important role assigned to these models in Tang tomb arrangements and their significance as status symbols and powerful guardians protecting the dead meant that these clay figures became luxury objects. No expense was spared in their production, and the larger examples are masterpieces of technology made from the finest clays and rare pigments. Created during one of the greatest periods in Chinese history, they reflect the artistic vitality of the time and give a unique glimpse into the luxurious and sophisticated world of contemporary upper class life.

Techniques and Production Methods

Technically the most important development during the

Tang was in the use of *sancai* or three-colour glaze. The Northern Wei had experimented with polychrome glazing, but its full potential could only be realized after the Sui and early Tang had developed a fine white-bodied ware which provided the perfect base for bright colours. By the end of the seventh century, the art of polychrome glazing had been perfected. By adding metal oxides to the lead glaze, the Tang produced a new range of colours remarkable for their intensity. The addition of copper produced green; cobalt gave deep Prussian blue; varying proportions of iron produced a wide range of shades from rich amber to soft yellow. The colours were applied by dripping, spotting, or with the use of wax resist (a method whereby certain parts of the body were protected from the glaze by wax). When fired, the colours ran, blurring and mixing to make fused colours. The fluid, irregular patterns which this method produced were intentional. Colouring enhanced the value of an object and did not need to be accurate. A blue horse was doubly prized: the horse, itself a symbol of rank, was given added value by its prestigious colour since cobalt, first imported from the Middle East in the early seventh century, was more valuable than gold.

The use of *sancai*, begun at the end of the seventh century, lasted until the mid-eighth, and very few examples have been found in later tombs. Production was centred in certain kilns — Yaozhou, near Tongchuan in Shaanxi, Neiqiu in Hebei, and Gongxian in Henan — and it is possible that these were damaged during the An Lushan rebellion.

Sancai was almost exclusively produced for tomb use; only a very few examples have been found among export wares. The expense of the long production process and costly minerals meant that *sancai* figures became highly prized status objects, reserved for but not always used by the upper classes. In some large tombs, all the figurines are unglazed;

in others, glazed and unglazed are found together. Certain parts of a glazed figure were left unglazed, including faces, some animal guardian crests which were painted, and the bases for horses and camels. Unglazed figures were covered with white slip and then painted with mineral pigments. According to contemporary records, entertainers, performers of ritual music and martial arts, and the imperial ceremonial troops often wore real gold and silver ornaments or armour. Traces of a gold leaf necklace have survived on the dancing boy now housed in the Victoria and Albert Museum, and 113 earthenware figures of horsemen found in an early Tang royal tomb have gold leaf frontals on their horses.

The figures were made with the classical mass production process using a mixture of moulded and modelled pieces. The separately moulded parts of the body were joined together by luting (applying a diluted clay mixture as glue), and the joins were then smoothed with a knife or by hand. Heads from the same mould could be made to look different by setting them on the body at different angles. In the larger horses and camels, one or occasionally two legs were solid to support the weight, and holes were left in the bellies and mouths of horses to allow steam to escape during firing. In some human figures, such as the dancing boys, there is a hole under one arm covered by a shawl across the shoulders. An armature of wood and iron may have been used to hold the larger pieces during the assembly process. Small modelled parts such as bridles, trappings, noses, ears, and belt buckles were then added, and finally the fine details incised with a knife. After a first firing at a low temperature, the glazes were added and the figure re-fired still at a relatively low temperature. The figures are therefore fragile, and with their weight must have caused considerable problems during transport.

5

Epilogue

FOR a long time it was believed that the Tang were the last to use tomb figurines. The practice of replacing figurines with paper models which were burned at the tomb, transforming them into smoke, a medium easily acceptable to the spirit, had begun in the late Tang period. Under the Song (960–1279), this practice was encouraged by widespread use of cremation. Marco Polo describes seeing paper models of houses, servants, animals, clothes, and even tomb figurines being burned with the body to satisfy the needs of the deceased after death. However, recent excavations have shown that although figurines no longer formed a standard part of tomb furnishings, the tradition lingered on. Finds are scattered geographically, and the decision whether or not to use figurines seems to have been a question of individual choice or local habit.

Song tombs were modest in size, and the number of figurines in a single tomb was small. Song subjects were almost exclusively confined to human figures created to serve the dead. Figurines from tombs in the southern part of the country are often of fine white porcelain with the newly developed *qingbai* glaze — an extremely thin blue-white glaze — over incised decoration. Small *qingbai* figures, some 20 centimetres high, from the kilns at Jingdezhen, Jiangsu, have been found of soldiers, officials, buildings, and gardens. In the northern areas, figurines were mostly uncoloured or painted biscuit. Male and female attendants in painted earthenware from a late Northern Song or Jin (1115–1234) tomb at Xinlifeng, Jiaozuo, Henan, wear elaborate headdresses and stand on ornamental bases, carrying household objects such as boxes, basins, or tea trays. The most

5.1 Actors, Jin Dynasty, twelfth century, Henan.

lively post-Tang group yet discovered is from a Jin tomb
in the same area, strengthening the belief this was a local
centre of production. They are a troupe of *zaju* players.
(*Zaju* was a form of mixed drama with singing, music,
dance, and comedy which was very popular in the twelfth
and thirteenth centuries.) The small figures are filled with
movement, caught in the act of whistling, clapping, or dancing (Fig. 5.1). Unusually, they are not modelled from wet
clay but chiselled from already baked bricks.

Figurines from tombs of the Yuan Dynasty (1271–1368)
tend to be of grey unglazed earthenware and reflect daily
life among northern peoples. Yuan tombs in Shaanxi have
yielded a wide variety of subjects: men and women in northern dress, carts, packhorses, mounted and unmounted horses and camels, domestic objects such as stoves, pots, bowls,
lamps, inkslabs, and even incense burners. Although carved

56

5.2 Folding chair with footstool tucked beneath it, glazed earthenware, early seventeenth century. Height 16.8 cm. Courtesy of the Glasgow Museums: The Burrell Collection.

with great detail, these small figures, seldom over 50 centimetres high, lack the movement of Tang figurines.

The Ming (1368–1644) consciously revived the Tang use of coloured glazes but with a different range of colours — light and dark blue, green, yellow, and turquoise. Human figures are on the whole small, without movement, usually representing officials or ladies of rank rather than servants. Reflecting contemporary interest in furniture, there are glazed models of tables, folding armchairs, various types of cupboards, and bookcases (Fig. 5.2). There are military guardians, once again based on human beings rather than deities, and miniature trays with highly decorated dishes of coloured fruit, fish, and kebabs, like wax models in a child's doll's house. Although the total number of figurines in any tomb remained modest, there was a penchant for

5.3 Two-courtyard house, fifteenth–sixteenth century. Height 44 cm. Courtesy of the Trustees of the British Museum.

complex groups — funeral processions with musicians or large two- or three-courtyard mansions. All the details of a Ming house are reproduced in miniature — the ridged green roof tiles, roof animals guarding the eaves, wall decorations, and the screen door which blocked the view into the courtyard from the street (Fig. 5.3). In a three-courtyard house representing a princely dwelling, guests are clearly expected. Servants stand beside tables laden with food and chopsticks in the first, most public courtyard, whilst the hosts stand ready to receive their guests in the second courtyard (Plate 20).

The number of excavated post-Tang figurines is still so few that it is difficult to discern any pattern in their use. On the available evidence, it looks as if the Song continued the Tang idea of providing the deceased with attendants, and regarded the figures, at least in the south, as giving status. In the Jin and Yuan dynasties which followed, the emphasis was on recreating daily life. With the Ming, the original thread seems to have broken. Although the houses are masterpieces of architectural reconstruction, as so often happens with a conscious revival of earlier fashions, the glazed figures lack spirit, and on the present scanty evidence, it is hard to see how they fit into the overall pattern of Ming beliefs about the tomb.

The real history of Chinese tomb figurines thus spans a millennium. From the clay figurine armies of the Han to the magnificent glazed models of the Tang, these miniatures reflect the changes in Chinese society. History is recreated below ground, and these models bring the past alive in a way that written texts can never do. With the exception of a short period in the Northern Dynasties, tomb figurines belong to the mainstream of the Chinese classical tradition of sculpture. The sculptors carved from life, catching the spirit of their age with an often devasting simplicity and directness. When the beliefs behind their use faded, the figurines lost their importance, and the habit of recreating a world below ground died away.

Chronological Table

Shang	*c.*1550–1027 BC
Western Zhou	1027–771 BC
Eastern Zhou	
Spring and Autumn period	770–475 BC
Warring States period	475–221 BC
Qin	221–207 BC
Western Han	206 BC–AD 9
Wang Mang	9–25
Eastern Han	25–220
Period of Disunity	
Three Kingdoms	221–280
Western Jin	265–316
Eastern Jin	317–420
Southern Dynasties	420–589
Northern Dynasties	
Northern Wei	386–534
Eastern Wei	534–550
Western Wei	535–557
Northern Qi	550–577
Northern Zhou	557–581
Sui	581–618
Tang	618–907
Five Dynasties	907–960
Song	960–1279
Jin	1115–1234
Yuan (Mongol)	1271–1368
Ming	1368–1644
Qing	1644–1911
Republic of China	1912–1949
People's Republic of China	1949–

Selected Bibliography

Fong, Mary H., 'Antecedents of Sui-Tang Burial Practices in Shaanxi', *Artibus Asiae*, LI, 3/4 (1991).

Jacobsen, Robert D., 'Ceramic Tomb Sets of Early T'ang', *The Minneapolis Institute of Arts Bulletin*, LXIV (1978–80), pp. 4–23.

Juliano, Annette, 'Teng-Hsien, An Important Six Dynasties Tomb', *Artibus, Asiae*, Supplement 37 (1980).

Kuwayama, George, ed., *Ancient Mortuary Traditions of China* (California, Los Angeles County Museum of Art and University of Hawaii Press, 1991).

Hentze, Carl, *Chinese Tomb Figures* (London, Goldston, 1928).

Homage to Heaven, Homage to Earth (Toronto, Royal Ontario Museum, 1992).

Hsu Chun, *L'évolution des* yong *statuettes funéraires de l'époque des Royaume Combattants aux Han Occidentaux* (Paris, 1987).

Kerr, Rose, ed., *Chinese Art and Design* (London, Victoria and Albert Museum, 1991).

Kesner, Ladislav, 'Portrait Aspects and Social Functions of Chinese Ceramic Tomb Sculpture', *Orientations*, 8 (1991), pp. 33–42.

Laufer, Berthold, 'Chinese Clay Figures', *Chicago Natural History Museum, Anthropological Series*, 13, 2 (1914).

Lewis, Candace J., *Into the Afterlife: Han and Six Dynasties Chinese Tomb Sculpture from the Schloss Collection* (New York, Vassar College Art Gallery, 1990).

Los Angeles County Museum of Art, *The Quest for Eternity: Chinese Ceramic Sculpture from the People's Republic of China* (Los Angeles, 1987).

Mahler, Jane, *Westerners among the Figurines of the T'ang Dynasty of China* (Rome, Istituto Italiano per il Medio ed Estremo Oriente, 1959).

Medley, Margaret, *The Chinese Potter* (Oxford, Phaidon Press, 3rd. ed., 1989).

Rawson, Jessica, ed., *The British Museum Book of Chinese Art* (London, British Museum Press, 1992).

Schloss, Ezekiel, *Chinese Pottery Figurines* (New York, Friendly House Publishers, 1963).

——, *Ancient Chinese Ceramic Sculpture from Han to T'ang*, 2 vols. (Stamford, Castle Publishing, 1977).

Till, Barry and Paula Swart, *Images from the Tomb: Chinese Burial Figurines* (Victoria, Art Gallery of Greater Victoria, 1988).

Vainker, Sheilagh, *Chinese Pottery and Porcelain: From Prehistory to the Present* (London, British Museum, 1991).

Index

CHINA

XINJIANG UYGUR
Autonomous Region

QINGHAI

TIBET
Autonomous Region